FIRST EXPERIMENTS IN SCIENCE & NATURE

Ecology and Pollution/Land

BY MARTIN J. GUTNIK

ILLUSTRATED BY JIM CARLETON

CHILDRENS PRESS, CHICAGO

Library of Congress Cataloging in Publication Data

Gutnik, Martin J
    Ecology and pollution/land.

    (First experiments in science and nature)
    SUMMARY: Instructions for experiments illustrating
the characteristics of a land ecosystem, land pollution,
and how waste materials can be used to benefit the land.
    1. Ecology—Experiments—Juvenile literature.
2. Pollution—Experiments—Juvenile literature.
[1. Ecology—Experiments. 2. Pollution—Experiments]
I. Carleton, Jim, illus. II. Title.
QH541.14.G88   628′.44   73-7795
ISBN 0-516-00523-5

3 4 5 6 7 8 9 10 11 12 13 14 15 16 17 18 19 20 21 R 79 78 77 76

*To Timothy Stearns
who is truly dedicated
to the study of ecology.*

CONTENTS

FOREWORD

Ecology is the study of how living things depend on each other and on their environment (the air, land, and water where they live).

An ecosystem is a place where certain types of plants and animals live together.

The ecology experiments in this book will show you how plants and animals live and work together in a land ecosystem. When you watch the ecosystem you build you will learn that every living thing has a purpose and exists so that other living things can continue to exist. They need and depend on each other.

The experiments in the pollution sections of this book will show you how man can put his waste to use instead of allowing the waste to pollute our land.

BUILDING A LAND ECOSYSTEM

Tim wanted to see how plants and animals live together. His teacher told him how to build a land ecosystem. He needed these things:

1. A ten-gallon fish tank.
2. A screen to fit over the top of the fish tank.
3. Enough soil to cover the bottom of the fish tank four inches deep. (He can get this from the ground near his house or from a place that sells soil for plants.)
4. A plastic dish about three inches deep.

5. A handful of grass seed.

6. Five or six plants.

7. About twelve earthworms.

8. Ten crickets. Tim knew he must not get grasshoppers. They would eat all the plants.

9. One frog or salamander or lizard.

10. A few small rocks and stones.

11. A light to go over the ecosystem.

12. Fruit flies or houseflies.

Tim Builds His Ecosystem

Tim put a plastic dish in one corner of
his fish tank. This would be his pond.
Then he spread the soil over the bottom
of the fish tank.

Now Tim was ready to plant the plants.
When he decided where he wanted to
put them, he dug the holes. He made
holes just deep enough to leave room
for the roots of the plants. Then Tim put
the plants in the holes and covered the
roots with soil.

He then spread some grass seed over
the ground where there were no plants.

FROG
CRICKETS

Tim used a pitcher to water the grass
and plants in his ecosystem. He watered
until puddles appeared on the soil.

Tim had to wait about one week before
he put the insects and animals into his
fish tank ecosystem. This gave the plants
a chance to grow.

After one week he put in the insects
and animals. He quickly put the screen
over the top of the fish tank. He put a
light over the fish tank. The light would
give both heat and light to Tim's
ecosystem.

How Tim Kept His Ecosystem Going

Tim filled the pond every day. He did this to be sure that the animals had fresh drinking water.

At least once a week Tim used his sprinkling can to water the ecosystem. He used his finger to help him decide whether or not to water it. If the soil felt moist, he didn't water. If it felt dry, he watered. He watered until the soil would take no more water. Puddles formed on the ground.

Each time Tim watered the ecosystem, he used his pitcher to change the water in the pond.

Tim found out what his insects and animals eat. (See appendix, page 42.) He made sure that they always had enough food.

His teacher told him that a
temperature of seventy-five to eighty
degrees Fahrenheit would be best for
his ecosystem. Frogs, salamanders, and
lizards are cold blooded. They won't eat
unless the temperature is kept this high.
Tim always kept his ecosystem at that
temperature. He did this by using 100-
watt light bulbs. He put the bulbs in
lights over the ecosystem. When he
wanted more heat he added more bulbs.

What Tim Saw

Tim kept a daily record of his ecosystem. On this record he put these things:

1. How the plants and grass were doing: Were they growing? He used a ruler to measure them.

2. How the insects were doing: Were they eating? What were they eating?

3. How the animals were doing: Were they eating? What were they eating? Were they growing? Which animals were eating plants? Which animals were eating insects or other animals?

DAILY RECORD OF LAND ECOSYSTEM

DATE	INSECTS-ANIMALS	DATE	PLANTS
3/13	CRICKETS EATING THE GRASS. FROG IS GROWING— ATE ONE CRICKET.		GRASS IS TWO INCHES HIGH. BEGONIA PLANT IS DYING.
3/14	BOUGHT TWO MORE CRICKETS. NOW HAVE ELEVEN IN ALL IN ECOSYSTEM. FRUIT FLIES ARE DOING WELL.		REMOVED DEAD BEGONIA PLANT—AND PUT IN IVY. TRIMMED GRASS WITH SCISSORS.
	CHANGED LIGHT BULB THAT BURNED OUT— 100 WATT.	3/15	WATERED ECOSYSTEM TODAY. EVERYTHING IS GROWING WELL!

BUILDING A MOISTURE GRADIENT BOX

(WATER BOX)

Tim's teacher told him that some
plants live in dry places, some plants
live in wet places, and some plants live in
places where it is not too wet or too dry.
He wanted to find out how different
kinds of plants would grow with
different amounts of water. He decided to
build a moisture gradient box.

Tim needed these things to build his moisture gradient box:

1. A large shoe box.
2. A large plastic bag. (A trash can liner would work well.)
3. Enough soil to fill the shoe box half full. Tim planned to get the soil from his yard.
4. Nine plants:

   a. Three "dry" plants (plants that live in dry areas). A cactus is a dry plant.

   b. Three "wet" plants (plants that live in water). An elodea is a wet plant. Tim could get these in a pet store.

   c. Three "middle" plants (plants that live where neither wet plants nor dry plants will grow). A coleus is a middle plant.
5. A piece of wood twelve inches high and six inches wide.
6. Masking tape.
7. A pitcher.

Tim Builds His Moisture Gradient Box

Tim put a large trash can liner inside
a shoe box. The edges of the plastic hung
over the sides of the shoe box.

He poured the soil into the shoe box
and spread it evenly over the bottom.
He made sure that the box was half full
of soil.

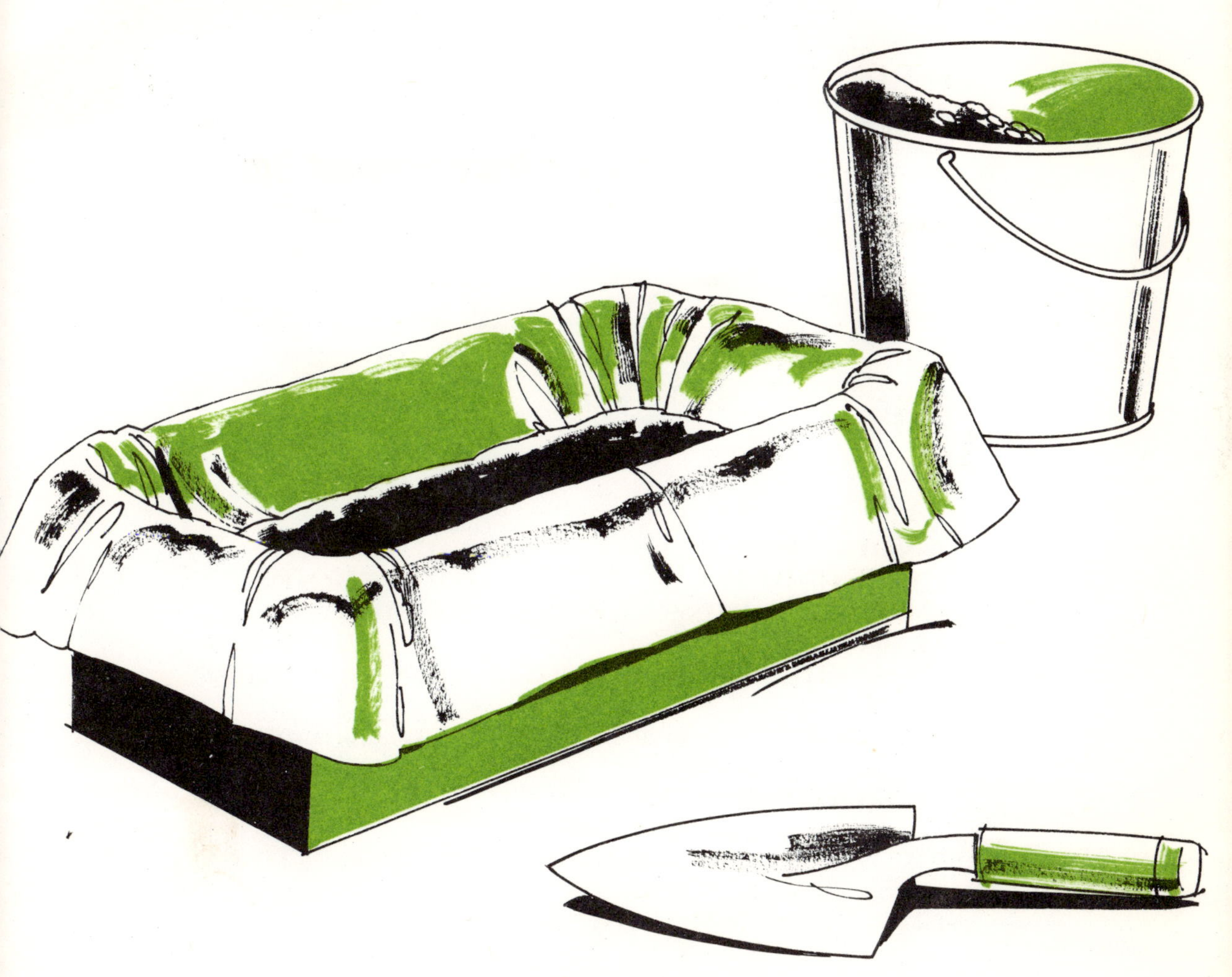

Now Tim was ready to put in the
plants. He dug a hole in the soil for
each plant. He made three rows of holes.
He made one row at the top of the box,
one row in the middle of the box, and
one row at the bottom of the box. There
were three holes in each row.

Tim put the dry plants in the holes
on the right side of the box. He covered
the roots with soil. He planted the wet
plants in the holes on the left side of the
box. He put the middle plants in the
middle row.

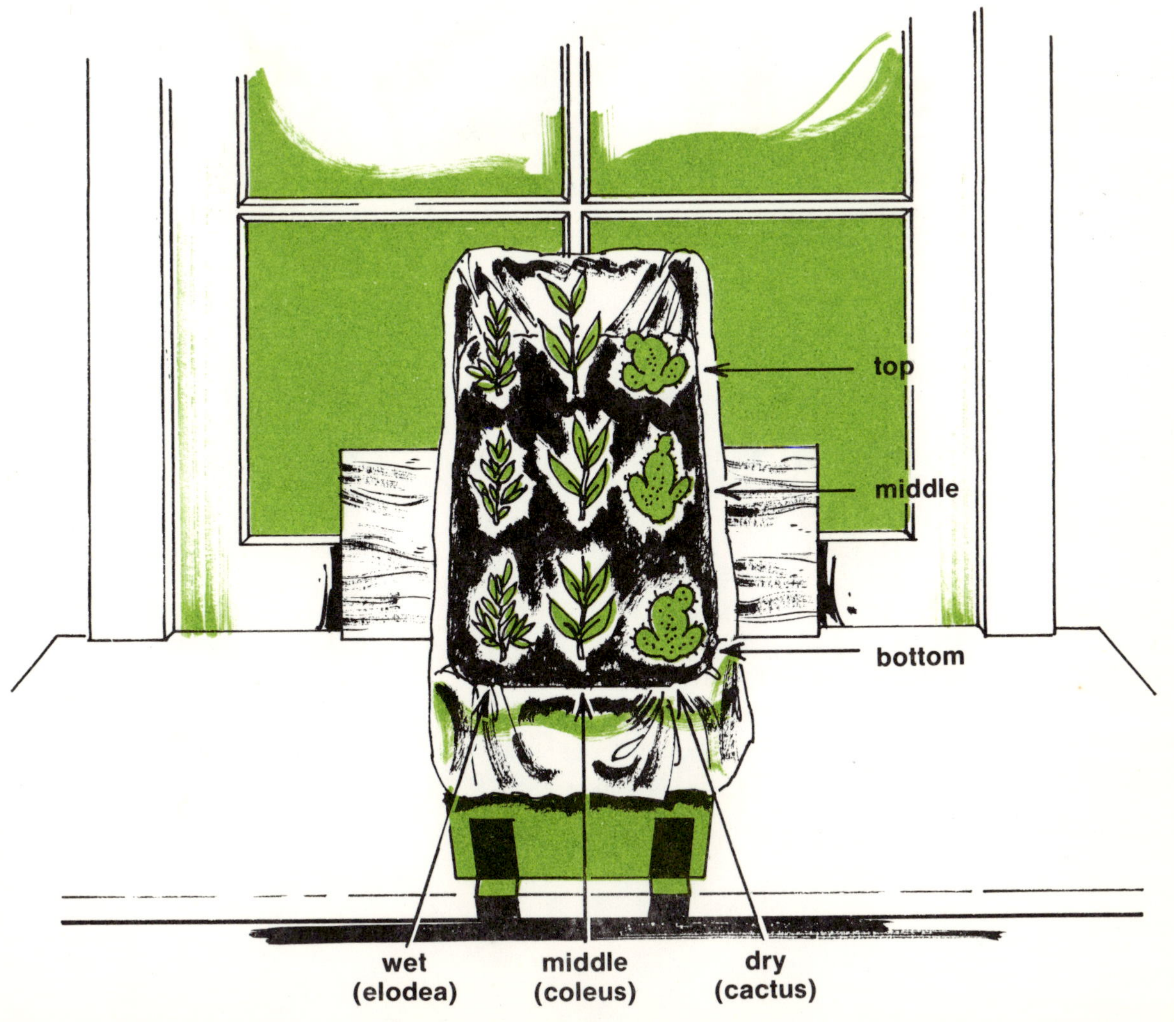

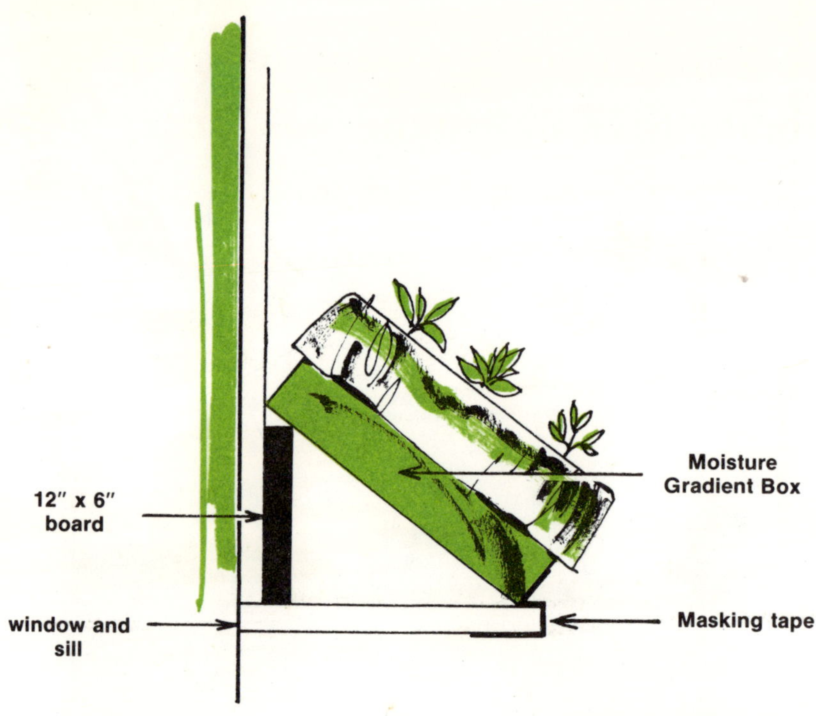

Tim's teacher told him that his water
box should be put near a window so the
plants would get enough light. He taped
his board to a window and set the box
against the board. He taped the box so it
wouldn't move. It looked like a slide. The
board was the back and the box was
slanted like a slide.

Tim used his pitcher to fill the
bottom of the box with water. He kept
putting water in the box until a pond
formed. He covered the bottom row of
plants with water. The water sank into
the soil.

Tim kept filling the bottom of the
box until the water stayed in a pond. *He
did not water the plants above the bottom
row. He only filled the pond.* Tim would
have to fill the pond once a day.

What Did Tim See?

Tim watched the water box every day. In one or two weeks he saw some changes.

The cactus plants (dry plants) were on the right side of the box. The one in the top row felt harder than the one in the middle row or the one in the bottom row.

Tim pulled out the cactus from the top row. He looked at its roots. They were firm and dry. The roots of the other two cactus plants were wet and soggy. They were dying. The cactus from the top row was much healthier than the other two cactus plants.

The elodea (wet plants) were on the
left side of the box. The one in the pond
at the bottom of the box was green,
firm, and healthy. The elodea plants at
the top and in the middle were dried up
and dead. They crumbled when Tim
held them in his hand.

The coleus (middle plants) were in the
middle of the box. The coleus in the
middle was firm, green, and healthy.
When Tim bent it over it popped back
up. The coleus in the top row was soft
and was losing its color. The one in the
bottom row was soggy and soft and was
losing its color.

Tim saw that dry plants do best in a
dry area; wet plants do best in a wet
area; and middle plants do best where it
is not too dry and not too wet.

HOW TO MAKE
A SANITARY LAND-FILL SITE

Because there are so many people on our earth, there is a lot of waste. People use many things. What they don't use, or what is left over, is waste. What does man do with all his waste? He has to put it somewhere. When waste is used in the wrong ways or put in the wrong places, it causes land pollution.

Tim wanted to know what people do with their waste. He decided to make a sanitary land-fill site in his land ecosystem. Tim needed these things for his experiment:

1. His land ecosystem.
2. One cup of charcoal; this can be bought at most pet stores.
3. One piece of paper.
4. A sprinkling can.
5. A small hand spade (trowel).

Tim Builds His Land-Fill Site

Tim took the screen off the top of his land ecosystem. He dug a hole with his hand spade. The hole was about four inches deep and six inches wide.

Tim tore his paper into tiny pieces. He mixed the paper with the charcoal. He put the paper and charcoal mixture into the hole.

He covered the paper and charcoal
with the soil he had dug to make the
hole. When Tim was through covering
up the paper and charcoal with the soil,
there was a little hill. He had made a
land-fill site. The paper and charcoal
were the waste.

Tim used his pitcher to water the
land-fill site. He watered it until the
water soaked into the hill. He watered
the site once a day for the next two weeks.

What Did Tim Look For?

Tim looked at the land-fill site every
day. He asked himself these questions:
What happens to the land-fill site when I
water it? Does the water push any soil
away? Does the water soak into the site?
Does the water go any deeper than the
site?

What Happened?

The water did push some of the soil off
the hill. The water did sink into the
land fill. The water also sank through the
land fill.

What Did This Show Tim?

This showed Tim how some of our
soil gets washed away (eroded). When it
rains on bare soil, the soil is washed
away with the water.

This experiment also showed Tim
how land-fill sites can cause pollution.
Rainwater sinks through land fills.
When it does this, the water gets dirty.
This dirty water then goes into wells
and underground rivers.

HOW TO MAKE PAPER
BY RECYCLING OLD PAPER

Man pollutes his world in many ways. One way is by wasting things that could be used again and again. When we throw things away instead of finding ways to use them again we are sometimes wasting. The waste causes pollution. It also uses up valuable resources that are hard to replace.

Paper is made from trees. If we keep on wasting paper, we will use up more and more of our trees. It takes many years for a tree to grow to replace a tree that has been used to make paper. We should use our paper again and again instead of throwing it away.

Tim wondered if it would be possible
to re-use paper. His teacher told him how
to do an experiment in recycling paper.
Tim needed these things for his
experiment:

1. White facial tissues. (This experiment
   must be done with white paper only.
   Special chemicals are needed for
   paper with colors or print.)
2. An egg beater.
3. A bowl.
4. A cake pan 13" x 9" or larger.
5. A piece of screen to fit in the cake
   pan.
6. Water.

Tim Does His Experiment

Tim put about thirty pieces of facial tissue into a bowl. He filled the bowl half full of warm water. He let the paper soak in the water for an hour.

After one hour Tim was ready to mix
the paper. He used the egg beater to beat
the paper until it got soupy.

This soupy mixture is called pulp.

Tim put his screen into the pan. He
poured the pulp into the pan. He lifted
up the screen and shook it gently over
the pan until there was a layer of pulp on
top of the screen.

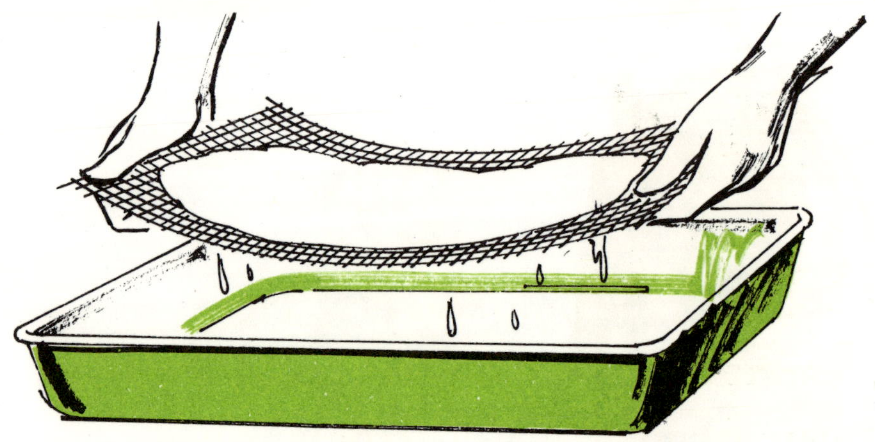

Tim lifted the screen out of the pan of
pulp and set it down to dry. He let it dry
for one day.

When the pulp was dry, Tim
carefully lifted it off the screen. He was
careful not to tear it. Tim now had
recycled paper.

What Did this Show Tim?

Tim saw that it is possible to stop
wasting paper. He knew that if we recycle
paper many trees would be saved.

Tim thought about the many other
things that can be recycled.

APPENDIX: WHAT THE INSECTS AND ANIMALS IN TIM'S LAND ECOSYSTEM EAT

Fruit flies: Ripe, squashed bananas.

Houseflies: Raw meat that is left to rot.

Crickets: Bananas and part of the plants in the ecosystem.

Frogs: Fruit flies, worms, and crickets.

Salamanders: Worms.

Lizards: Worms, flies, and crickets.

WORDS YOU SHOULD KNOW

Ecology (ee KAHL uh gee)   The study of how
living things depend on each other
and on their environment (the
air, land, and water where they live).

Ecosystem (EE ko sis tem)   A place where
certain types of plants and animals
live together. A pond is an
ecosystem. A forest is an ecosystem.

Environment (ehn VYE ruhn ment)
Surroundings that affect plants and
animals; the air, land, and water
where they live.

Eroded (ih ROH dehd)   Worn away or
washed away. Land is eroded when
the topsoil has been worn away or
washed away.

Land fill (land fill)   A place where garbage
or trash is dumped and then
covered over with earth.

Recycle (ree SYE k'l)   To use over again.

BIBLIOGRAPY OF RESOURCES

Barnard, Philip: Fourth-grade teacher, Atwater
Elementary School, Shorewood, Wisconsin.
Stearns, Forest: Professor of Ecology, University
of Wisconsin, Milwaukee, Wisconsin.
Stearns, Timothy: Fifth-grade pupil, Atwater
Elementary School, Shorewood, Wisconsin.
Tranetski, John: Graduate Student in Botany,
University of Wisconsin, Milwaukee Wisconsin.

About the Author: Martin Gutnik, an innovative elementary school science teacher, lives with his wife and two young children in Milwaukee, Wisconsin. Because he was unable to find suitable books of experiments in ecology and pollution, Mr. Gutnik created his own experiments for his students to perform in the classroom. He has set these experiments down in the first three books of his new series, *First Experiments in Science and Nature,* so that children everywhere will be able to discover and enjoy science on their own. Though Mr. Gutnik spends much of his time helping his students learn to build ecosystems, develop film, and dissect frogs, his life is not totally surrounded by science. He also collects records from the fifties and enjoys singing folk songs as he accompanies himself on the guitar, which he taught himself to play.

About the Artist: Jim Carleton attended Chicago's American Academy of Art as well as the Art Institute, and has lived most of his life in the Chicago area. His main work is done in publishing and advertising. When he is not involved in art, his favorite pastimes are bicycle riding, tent camping and fishing. Jim also takes great pleasure in cooking his catch in the open air on camping expeditions. He lives with his wife and five children in suburban Villa Park, Illinois.